THE POETRY OF RUTHERFORDIUM

The Poetry of Rutherfordium

Walter the Educator

Silent King Books

SILENT KING BOOKS

SKB

Copyright © 2024 by Walter the Educator

All rights reserved. No part of this book may be reproduced in any manner whatsoever without written permission except in the case of brief quotations embodied in critical articles and reviews.

First Printing, 2024

Disclaimer
This book is a literary work; poems are not about specific persons, locations, situations, and/or circumstances unless mentioned in a historical context. This book is for entertainment and informational purposes only. The author and publisher offer this information without warranties expressed or implied. No matter the grounds, neither the author nor the publisher will be accountable for any losses, injuries, or other damages caused by the reader's use of this book. The use of this book acknowledges an understanding and acceptance of this disclaimer.

"Earning a degree in chemistry changed my life!"
— Walter the Educator

dedicated to all the chemistry lovers, like myself, across the world

RUTHERFORDIUM

Of elements, rare and profound,

RUTHERFORDIUM

Resides a titan, Rutherfordium, renowned.

RUTHERFORDIUM

With atomic number 104, it claims its throne,

RUTHERFORDIUM

In the periodic table's kingdom, all its own.

RUTHERFORDIUM

Born from fusion in the hearts of stars,

RUTHERFORDIUM

Its journey through cosmos, filled with memoirs.

RUTHERFORDIUM

Synthesized by human hands, in labs it gleams,

RUTHERFORDIUM

Unveiling mysteries, beyond our wildest dreams.

RUTHERFORDIUM

In the nucleus, a nucleus, dense and compact,

RUTHERFORDIUM

Protons and neutrons, in a delicate pact.

RUTHERFORDIUM

Yet unstable, it yearns to break free,

RUTHERFORDIUM

Into particles, it disintegrates with glee.

RUTHERFORDIUM

A fleeting existence, a transient dance,

RUTHERFORDIUM

In the cosmic theater, it takes its chance.

RUTHERFORDIUM

For mere moments, it graces our sight,

RUTHERFORDIUM

Before fading away, into the endless night.

RUTHERFORDIUM

But in its brief existence, it leaves its mark,

RUTHERFORDIUM

In the annals of science, a shining spark.

RUTHERFORDIUM

A testament to human curiosity and endeavor,

RUTHERFORDIUM

Unraveling nature's secrets, now and forever

RUTHERFORDIUM

Its name pays homage, to Ernest Rutherford's might,

RUTHERFORDIUM

A pioneer of atoms, whose genius took flight.

RUTHERFORDIUM

With daring experiments, he paved the way,

RUTHERFORDIUM

For the discovery of elements, in the grand display.

RUTHERFORDIUM

Rutherfordium, a symbol of human ingenuity,

RUTHERFORDIUM

Pushing the boundaries of possibility.

RUTHERFORDIUM

In laboratories, minds ignite,

RUTHERFORDIUM

As they harness its power, in the quest for light.

RUTHERFORDIUM

For within its core, lies potential untold,

RUTHERFORDIUM

To unlock mysteries, yet to unfold.

RUTHERFORDIUM

In the crucible of discovery, we strive,

RUTHERFORDIUM

To unravel the secrets, of this element's dive.

RUTHERFORDIUM

Yet in its legacy, it leaves a mark,

RUTHERFORDIUM

A testament to humanity's spark,

RUTHERFORDIUM

To push the boundaries of knowledge's reach,

RUTHERFORDIUM

In the depths of the atomic beach.

RUTHERFORDIUM

So let us marvel at Rutherfordium's glow,

RUTHERFORDIUM

In the periodic table's ebb and flow,

RUTHERFORDIUM

A symbol of human curiosity,

RUTHERFORDIUM

Unveiling the secrets of atomic diversity.

RUTHERFORDIUM

ABOUT THE CREATOR

Walter the Educator is one of the pseudonyms for Walter Anderson. Formally educated in Chemistry, Business, and Education, he is an educator, an author, a diverse entrepreneur, and he is the son of a disabled war veteran. "Walter the Educator" shares his time between educating and creating. He holds interests and owns several creative projects that entertain, enlighten, enhance, and educate, hoping to inspire and motivate you.

Follow, find new works, and stay up to date with Walter the Educator™ at WaltertheEducator.com

www.ingramcontent.com/pod-product-compliance
Lightning Source LLC
LaVergne TN
LVHW010412070526
838199LV00064B/5275